The Geology of Watkins Glen State Park:
A window into the Devonian past

By William A. Szary

Copyright 2019. Earth2Energy. All Rights Reserved.

Book Cover: Waterfalls in the Watkins Glen State Park is one of many that existed when glacial ice advanced and retreated during the Pleistocene. The falls shown are one of many along the park trail that can be viewed during the summer.

Library of Congress Catalog in Publications Data:

Szary, William A. The Geology of Watkins Glen State Park: A window into the Devonian past.

Includes references

ISBN 13: 9781707936731

Earth2Energy Educational Publishing
Port Richey FL 34668

Earth2Energy is a Registered Trademark

Table of Contents

Chapter 1. Geology of Western and Central New York 4
    An Overview

Chapter 2. Regional Geomorphology 10

Chapter 3. Physical Stratigraphy of the Genesee Formation (Devonian) In Western and Central New York 17
Introduction 17
Rocks Below the Genesee Formation 10
Genesee Formation 21
 Geneseo Shale Member
 Penn Yan Shale Member
 Sherburne Flagstone Member
 Renwick Shale Member
 Ithaca Member
 Genundewa Limestone Member
 West River Shale Member
Depositional Environment 38
Rocks Above the Genesee Formation: Sonyea Formation 40

Chapter 4. Watkins Glen State Park Geology 41
Park Geology 41
Bouma Sequence Discussion 42
 Introduction
 Flow Regime Indicators in Watkins Glen State Park

Chapter 1. Geology of Western and Central New York

An Overview

Guild (2011) posted a blog on the internet which summarized the geological history of Western and Central New York. Graphics were added to enhance the reading experience.

The geology of Western and Central New York tells an incredible story, laid out in the rolling hills, picturesque valleys, and the continuous road cuts throughout this region. The cascading waterfalls of Watkins Glen and Ithaca, along with the equally dramatic rock formations of Cattauraugus and Chautauqua counties, round out this state with an interesting preview into Devonian geologic history, 350 mya (**Figure 1**).

Figure 1. About 350 mya, Laurentia was submerged for the most part in the U.S except for most of central and northern New York State where the Appalachians were uplifting. A small portion of Western New York remained submerged beneath the Iaepetus Ocean. The Finger Lakes region was that area that was submerged.

Bedrock of the region was deposited over 380 mya in an equatorial continent that rotated about 45 degrees clockwise from its current position. Fossil remains and geological features indicate this area was submerged beneath a vast shallow ocean.

A continental collision with Baltica (proto Europe) began about 410 mya which uplifted the Acadian Mountains that was positioned along North America's east coast. The mountains were gradually eroded as the finer particles flowed deep out into the Catskill Delta which covered most of the Southern Tier of New York State, Pennsylvania, Ohio, West Virginia, and eastern Kentucky (**Figure 2**).

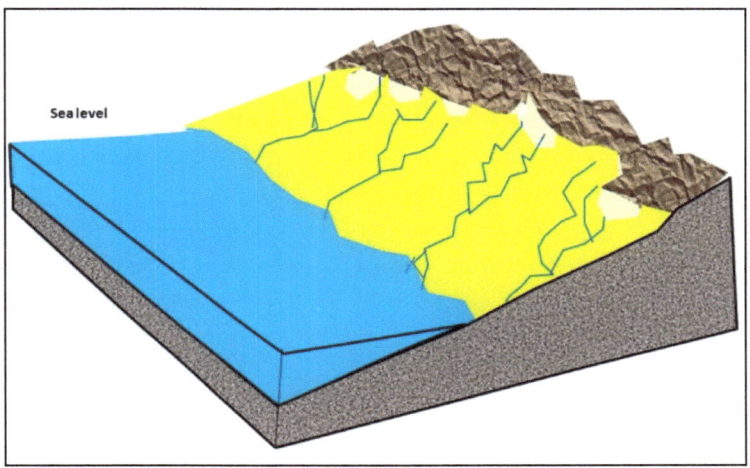

Figure 2. Model of the Catskill Delta forming along the shoreline of the Iaepetus Ocean as it covered Western New York.

Over vast periods of time, those particulates settled to the bottom of the ocean creating layer upon layer of sediment that eventually lithified. Clay sediments turned into shale, silt became siltstone, and sand compacted into sandstone (**Figure 3**).

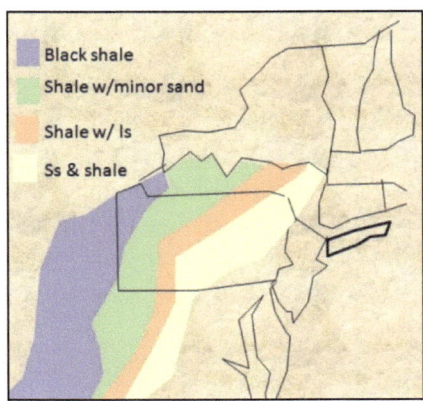

Figure 3. Northeastern U.S Devonian depositional pattern.

About 380 mya, the area was covered beneath a shallow sea. Climate hovered around 80 degrees F with no appreciable winter or snow. Fossils accumulated at the bottom of the shallow sea including brachiopods, crinoids, coral, placoderm fish, eurypterids, and trilobites. Mammals did not exist at this time nor did dinosaurs.

As the mountain sediment inflow slowly filled the sea and worldwide oceanic levels trended downward, the area slowly approached the shoreline of the shrinking ocean. Today's sandstone remains are evidence of the bottom of the shallow sea. Pebbles in quartz conglomerate were flushed into the shallow sea by torrential floods (**Figure 4**).

Figure 4. Quartz pebble conglomerate of the Rock City Formation near Olean, New York. Left is the outcrop exposure, right is the close up detail of the exposure. Source: Guild 2011.

The earth remained tectonically active when another continental collision occurred with Africa which uplifted the Allegheny Mountains between 360 and 345 mya. Pangea was formed, lasting between 330 and 220 mya. Dinosaurs dominated the landscape at this time. About 190 mya, the Atlantic Ocean opened and continued expanding. Laurentia was propelled northward by north to west plate tectonics, bringing it closer pushing it closer to its present day position. During this gradual transition, an asteroid struck the Yucatan Peninsula about 66 mya resulting in a cataclysmic disaster which wiped out the dinosaur population. Small rodent mammals began to thrive, giving rise to the mammal species and humans. The following series of paleo-tectonic maps provides a visual prospective on the position of the Laurentia from 330 to 60 mya (**Figures 5, & 6**).

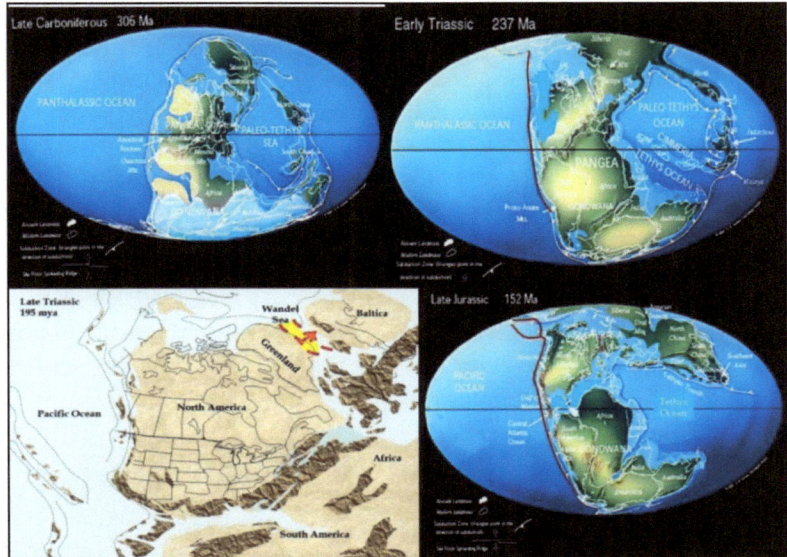

Figure 5. A glimpse into the positions of Laurentia during the Carboniferous at 306 mya (upper left); Early Triassic at 237 mya (upper right); and, Late Jurassic time, 152 mya (lower right). The Lower left map shows the approaching collision of Africa with Laurentia during the Allegheny Uplift during the Late Triassic Period 195 mya. Source of the Paleogeographic maps by C.R. Scotese, 2013 PALEOMAP Project, www.scotese.com.

Figure 6. Paleogeographic maps showing North America after Pangea broke up during the Middle Cretaceous (100 mya-upper left); Late Cretaceous (94 mya- upper right); Late Cretaceous (85 mya- lower left); and Early Paleocene (60 mya- lower right). Source of the Paleogeographic maps by C.R. Scotese, 2013 PALEOMAP Project, www.scotese.com.

About 1.8 mya, the geology of the region remained relatively stable. Huge and powerful glaciers carved their way through the landscape gouging out deep valleys and the Finger Lake crevasses. These ice sheets slowly plowed through those long accumulated and lithified deposits creating new lower grounds and substantial opportunities for erosion. The surviving Rock City Formation constitutes the locations in New York that remained unscoured by the powerful glaciers. Those areas that were not impacted by glacial ice were left exposed as the under layers of more delicate sedimentary rocks (shales) which led to the eventual erosion of the base material along with the characteristic breakage and shifting of conglomerate structures that produce the city like structures we see today (**Figure 7**).

Figure 7. Paleogeographic maps showing North America during Late Miocene (10 mya-left); and at 2 mya and 100 thousand years during the ice age-right).

The deep Finger Lake gouges left by the glaciers also left hanging falls that slowly eroded the formations exposed at Watkins Glen, Buttermilk Falls and Taugonnock Falls in Ithaca, New York (**Figures 8, 9, & 10**).

Figure 8. Upper Left- Rock City quartz conglomerate exposed at the base of the shale sequence in Watkins Glen State Park. Lower left- The Catherine River cut a deep gorge into the layers of shale in Watkins Glen. Waterfalls are abundant in the state park (right). Source: The right photo was posted on the internet. Rocks belong to the Sherburne Member of the Genesee Formation.

Figure 9. Buttermilk Falls, Ithaca, New York. Sources: Upper left photo posted on the internet by Imgur; Bottom left posted by Sanford Photography; Right photo posted on Wikipedia.

Figure 10. Taugannock Falls, Ithaca, New York. Left photo posted on the internet by Flickr; the right side was posted by Ithaca, NY and Scenic USA.

Chapter 2. Regional Geomorphology

Bloom (1965) provided a summary of the regional geomorphology from observations made 10 miles to the east of Seneca Lake where Watkins Glen State Park is located at the south end.

The Finger Lakes Region of Central New York is justly famous for two aspects of its geology: the Devonian stratigraphy and the Quaternary geomorphology. Not as well appreciated is the fact that the Quaternary landscape is the result of 360 million years of post-Devonian erosional history, for which no stratigraphic record is available in the region. An enormous unconformity everywhere separates glacial drift of late, or at the oldest, middle Pleistocene age, from the underlying lithified and mildly deformed Devonian marine strata. Volumes of sediment on the North Atlantic continental shelf and rise imply at least 2 km of regional denudation in the Cenozoic Era. The remaining Paleozoic section in the Finger Lakes region is little more than 2 km thick, so at least half of the depositional section is gone. With a regional southward dip of about 1 per cent, 2 km of vertical denudation involved 200 km of homoclinal shifting of the present north- facing escarpments. The landscape viewed from the hill tops around Ithaca has a grand story to tell, when we can learn to listen.

The key to the geomorphology of the Finger Lakes Region lies in the geometry of deposition, deformation, and denudation. The Upper Devonian sedimentary facies were deposited in an epicontinental sea with a rising source area to the east. Facies boundaries trend northeast to southwest, with Catskill facies alluvial plain red beds to the east, near shore marine sandstones next to the west, all of these grading westward into shales. Post-depositional regional deformation gently folded the Devonian rocks along fold axes trending N70E, significantly shortened the sedimentary pile in a N - S direction, and regionally tilted the pile southward about one-half degree (1 percent, 10 m/km, or 50 ft/mi).

The upper Paleozoic rocks of New York State record no significant source area to the north. The exposure of the Canadian Shield and Adirondack Highlands is therefore post-Devonian, although the former northward extent of Devonian sedimentation onto the continental platform is unknown. By Cenozoic time, the rivers of eastern North America were probably adjusted to structure on a regional scale. An inter-cuesta lowland (now Lake Ontario) should have followed the Ordovician shale belt between the shield and the Lockport dolomite escarpment. An ancestral Hudson River probably was eroding head ward along the Hudson Valley shale belt at the base of the Shawangunk and Catskill escarpments, and at its head ancestral Mohawk and Lake George lowlands probably were forming on shale belts broadly concentric to the Adirondack highlands. If the present is a key to the past, regional erosional denudation was accomplished primarily by homoclinal shifting of subsequent (strike - oriented) rivers along shale lowlands.

On the southern dip slopes of Upper Devonian shales, siltstones, and sandstones in central New York, regional dendritic south- flowing rivers can be reasonably inferred. River gradients should have been significantly less than the regional dip, so dipping strata would be truncated by erosion in progressively younger belts from north to south.
The highest area between Cayuga Lake and Seneca Lake (Connecticut Hill height 2099 ft) is a synclinal ridge, demonstrating topographic inversion of relief even on the gentle structures of the region. Several fossil peneplains have been proposed for the uplands of the Finger Lakes Region. The evidence is fragmentary, but a reductionist logic still permits: (1) a structure beveling ancient surface of low relief (and possibly near sea-level) now represented by the numerous summits in the region that range in height between 1800 and 2000 feet above sea level; (2) a lower structurally controlled surface primarily on shales at about 1000-1200 feet above sea level, and (3) various erosional levels within valleys below the two regionally correlative upland surfaces.

Ignoring several generations of scholarly research, we can suppose that the gentle regional uplift that created south-draining dip-slope river systems also initiated the subsequent river systems along shale belts and started a long history of homoclinal shifting and lateral migration of divides toward the south.

Ever since an early stage of regional erosion, north-flowing escarpment streams would have had steeper gradients than their south-flowing dip - slope counterparts, and the cuestas should have been migrating southward. Probably by late Cenozoic time, the region had evolved to a "broad valley" stage, with valley floors graded to a regional level now 900 – 1000 feet above sea level. Uplift (Cenozoic tectonism in New York?) rejuvenated the rivers, and a " deep stage" of in-trenched inner valleys resulted, especially along the axes of the north-flowing escarpment streams. These valleys became the precursors of the Finger Lakes. Presumably the rejuvenation caused more rapid divide migration toward the south. It has long been noted that barbed tributaries are common in the Finger Lakes, with acute junction angles of drainage systems pointing south but now draining north. Near Ithaca, both Salmon Creek and Fall Creek show this pattern; Keuka Lake is a Y- shaped glacially eroded lake basin that outlines a pair of southward-merging ancestral river valleys, although it now drains north to the St. Lawrence along with all the other Finger Lakes.

Glacial erosion and deposition has massively modified the ancestral fluvial systems that drain northward, converting north trending valleys into glacial troughs or rock basins many of which contain finger lakes. However, the glacial lowering of summit levels was minor, and the position of the regional St. Lawrence-Susquehanna divide may not have shifted very much by glacial processes. The evidence is weak, but the zone of highest summits across New York south of the Finger Lakes usually crosses the intervening valleys within a few miles of the present divide in each valley. The divide regions were reamed out by glacial-erosion and associated proglacial and ice-marginal melt water drainage, creating the famous "through valleys" that is the distinctive geomorphic feature of the region, much more so than the trough lakes which are common in glaciated regions. One brief published report concluded that the massive Valley Heads moraine at Tully, New York which forms the surface-water divide between Onondaga Creek to the north and the Tioughnioga system to the south by a barrier with 600 feet of relief on its northern proximal face may rest on a bedrock sill at a depth of only about 220 feet. The rivers that drain south from the divide flow on valley trains that are not deep over bedrock.

Their valleys were straightened by glacial erosion, but were not deepened very much. As most Valley Heads moraines in central New York crest at about 1200 - 1400 feet above sea level, the inferred bedrock sill beneath them may well occur at about the 1000 ft level that characterized the preglacial heads of the south-flowing drainage systems.

The hanging valleys and gorgeous gorges of the Finger Lakes Region are the result of glacial over-deepening of the main north draining valleys that had notched into the northern edge of the Allegheny Plateau in preglacial time. The lateral (east-west) tributaries lay athwart the dominant direction of ice flow and were subject more to glacial and glacio-fluvial deposition than to glacial erosion. The series of valleys that converge at Ithaca make the point very nicely: north-oriented Inlet Valley and the Cayuga Lake trough are eroded down to below sea level; northwest trending Sixmile Creek Valley and Cascadilla Creek Valley ("Ellis Hollow") have deep glacial lacustrine fills on their floors but expose bedrock in places at about 950 feet above sea level; Fall Creek, which flows west, has a few areas of exposed bedrock at about 900-950 feet above sea level but masses of glacial drift fill most of the preglacial valley. The depth of glacial erosion seems directly related to the degree that each valley funneled the advancing tongues of successive continental ice sheets (**Figure 11**).

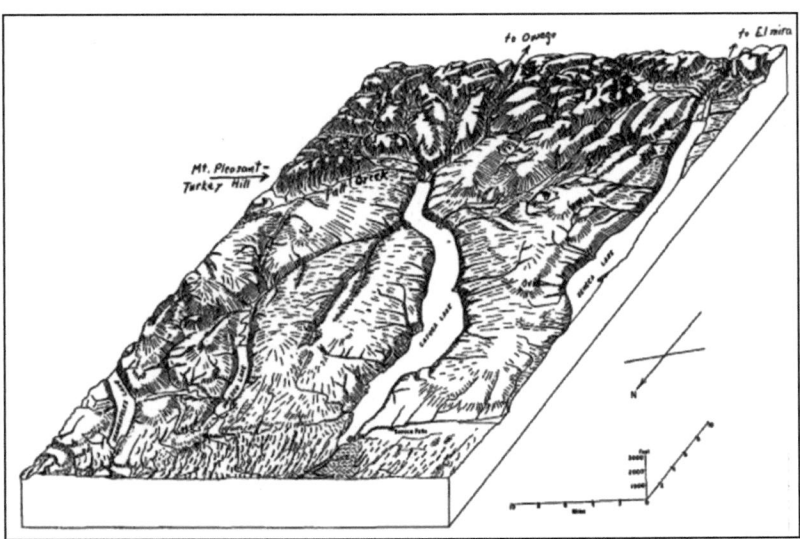

Figure 11. Physiographic diagram of the Cayuga Lake Basin. Source: Bloom 1965.

How many glaciers have covered Ithaca? From the deep sea oxygen isotope record, one can safely infer 20 ice ages of similar intensity to the last one, each lasting about 100,000 years. The classic 4–fold midwestern glacial subdivision of the Pleistocene is as obsolete as pre- plate tectonics land bridges and Atlantis, but the Finger Lakes Region has revealed little of its glacial past. A pre-Wisconsin valley fill at "Fernbank" on the west side of Cayuga Lake demonstrates that a gorge had been eroded previously, probably by a minor tributary that had incised the valley side after an earlier glacial event had deepened Cayuga trough. Therefore, at least two and possibly three glacial events can be inferred, depending on some assumptions. "Old" non-calcareous glacial drift is exposed at times along Sixmile Creek and at the head of Beebe Lake. If this drift is pre-Wisconsin, then the valleys around Ithaca were already as big as they are at the present prior to two ice ages ago. To the extent that their shape is due to glacial erosion, at least a previous third ice age must be inferred. The multi-reflecting sub-bottom seismic profile of Seneca Lake also implies that a long history of glacial erosion and deposition shaped the present catenary cross section of the Seneca Lake basin (north of Watkins Glen) (**Figure 12**).

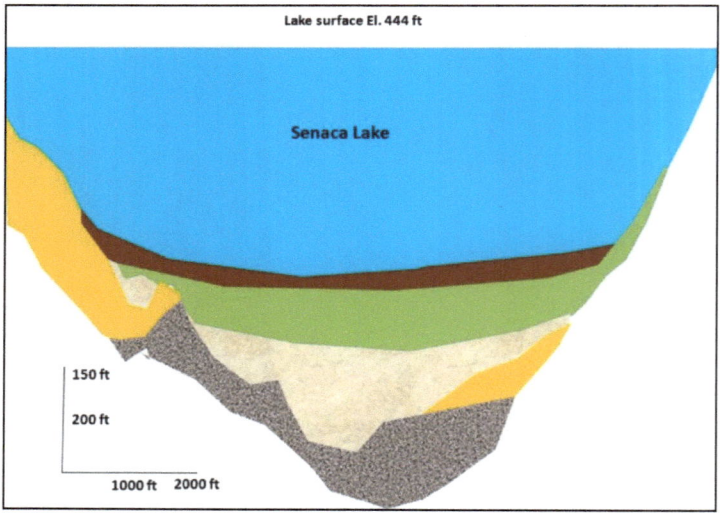

Figure 12. A seismic reflection profile across Seneca Lake based on data from a study for the Naval Research is 475 ft along this profile. Deep reflections are from sediment at least 500 feet below sea level. Vertical exaggeration X7.

As we approach the present, the record of geomorphic evolution of the Finger Lakes Region becomes more readable. Most of the tributary valleys of Cayuga Lake near Ithaca were aggraded by a variety of glacio-fluvial and glacial-lacustrine deposits as the Wisconsin ice sheet advanced southward against the regional drainage.

Rising preglacial lakes overflowed to the south across the divides, but most of the sediment remained within these great settling basins. Approximately 1000 annual couplets (varves) were counted in Sixmile Creek Valley, grouped in four series that progressively thicken upward and show an increase in the thickness of the winter layer. The height of these deposits above the Cayuga Valley floor suggests that an ice lobe completely blocked the north end of the lake; the 1000 layers, if annual, suggest that the ice advanced about 40 miles (65 km) in 1000 years or about 65 m/yr. A finite radiocarbon date of 41, 900 years from near the base of the varved sequence has been correlated with a post-Plum Point Interstadial ice advance. This date and several other "dead" radiocarbon analyses confirm that at least one mid-Wisconsin ice advance reached or affected the Finger Lakes Region.

The final push of Wisconsin ice crossed Ithaca and moved south to the vicinity of Williamsport, PA. The age of that terminal position is still debated, but 17,000 to 19,000 years will be an educated guess until southern colleagues agree on each other's evidence. The much favored relationship for the gradient at the edge of ice sheets, (h = ice thickness in meters, d = distance into the ice sheet from its margin, also in meters) yields a thickness of 1880 m of ice over Ithaca when the ice margin was 160 km to the south near Williamsport. Certainly, at this stage all the topography around Ithaca (maximum relief 2000 ft or 600 m) was deeply buried by ice.

By 13,000-14,000 years ago the ice margin had retreated to the regional divide south of the Finger Lakes. Here the ice edge paused or fluctuated, fed by a thickness of 1000 m or more of ice moving uphill toward the divide from the north, but unable to sustain glacier tongues down the Susquehanna system valleys. Here it built the Valley Heads moraine system, a series of massive morainal debris piles that choke the floors of the through valleys and with few exceptions determine the modern surface divide between the Susquehanna and St. Lawrence Rivers. With the progressive retreat of the ice margin northward from the Valley Heads moraine systems, the ice margin became more and more lobate and confined to the valleys. The South Cortland moraine, for example, was made by a distributary sublobe of the Cayuga Lake lobe that flowed northeastward for at least 12 miles from its base at Ithaca.

The outwash from the South Cortland moraine flowed northeast and east to Cortland, where it merged with the valley train down the Tioghnioga Valley from the Tully moraine.

Near-contemporaneity of several of the Valley Heads moraines in adjacent valleys can be demonstrated by similar relationships. Once north of the Valley Heads position, the retreating lobate ice margin terminated in preglacial lakes in all the valleys. All but the suspended fine silt and clay-sized sediment was trapped in the lakes. Notable around Ithaca is the pink or brown colors of the post Valley Heads lacustrine beds. The color is derived from the Silurian red shale belt across the north end of Cayuga Lake, fully 40 miles away. The reddish sediments contrast sharply with the local sediment derived from gray-colored source rocks, so the abandonment of a south-flowing melt water channel in favor of a lower western or eastern outlet, such as the rock gorges at Syracuse, is marked in the Ithaca region by an abrupt change from reddish to grayish sediments. The change is usually at a shallow depth, sometimes almost within the modern soil profile.

Deglaciation of the Cayuga Basin was even more rapid than earlier generations of geologists suspected. If the Valley Heads moraines are 13,000-14,000 years old and the St. Lawrence lowland was ice-free and open to the Champlain Sea by 12,000 years ago, less than 2000 years were required to deglaciate the northern two - thirds of New York State!

Some 10 or 15 names have been given to the sequence of hanging deltas in the Cayuga basin that record the episodic drop of proglacial lake levels, so each delta must have been built within a century or two. While the internal structure of the hanging deltas, with thick-bedded fore set gravel layers at angle of repose and only minor bottom set and subaerial top set beds, confirms rapid progradational growth, probably no geologist of the pre-radiocarbon era would have attributed their formation to a century time scale.

The river terraces and abandoned meanders along Fall Creek near Ithaca tell a similar story of rapid evolution. The oldest and highest surfaces record deposition or erosion in a variety of ice marginal streams or lakes, but with the integration of local lakes into glacial Lake Ithaca (outlet elevation 980 ft), lacustrine sediments and related hanging deltas and river terraces become reasonably correlative. A series of terraces near Varna record the dissection of the floor of local Freeville- Dryden Lake (elevation 1060 ft) by Fall Creek as the creek cut down to the level of glacial Lake Ithaca and lower levels. The terraces should project down-valley into the surfaces of the appropriate hanging deltas, but the uncertainty of the channel gradients and the rapidity of changing levels make detailed correlation unlikely.

The most significant fact is that all the terraces and hanging deltas were cut or built within one or two thousand years, at least 12,000 years ago. Soil profiles on the surfaces should differ more because of parent material, vegetation, and slope than because of climate or time. All the terraces surfaces around Ithaca probably had soils forming through the late-glacial and postglacial times of the tundra (?), spruce, pine, and hardwood forest succession.

As the late- glacial gorge entrenchment encountered buried bedrock spurs and former valley floors, further down cutting was severely inhibited. Postglacial gorge cutting has proceeded downward for a few tens of meters and head ward for some hundreds of meters, but most of our landscape has probably not changed much for 12,000 years. Even the impressive head ward retreat of Fall Creek at Ithaca Falls, Enfield Creek at Lucifer Falls, or Taughannock Creek at Taughannock Falls may have been accomplished primarily by the re - excavation of preexisting valley fills left by earlier glaciations. The landscape we see on excursions around Ithaca was shaped in a brief interval of intense activity and rapid change about 14,000-12,000 years ago. Mastodons and Paleo-Indians walked on the same landscape that we now stroll.

Chapter 3. Physical Stratigraphy of the Genesee Formation (Devonian) in Western and Central New York

Dewitt and Colton (1978) provided a summary of the physical stratigraphy in Western and Central New York describing the Genesee Formation and its members which are exposed in the canyon of Watkin Glen State Park. West of Cayuga Lake, the Genesee Formation thins by transgression on the Algonquin arch, as is shown by stratigraphic sections and conodont biostratigraphy.

Introduction

The area of this report covers parts of 14 counties in central and western New York and includes much of the Finger Lakes district, Genesee River valley, and a little of the south shore of Lake Erie. Western New York has long been a classic reference for the Upper Devonian sequence in North America. The strata in this area, particularly the lower Upper Devonian rocks of the Finger Lakes district, have been studied by many geologists during the past 130 years.

Knowledge of the stratigraphy in the area evolved slowly because of the scarcity of well-exposed sections and the complex interfingering of petroliferous brownish-black shale, calcareous gray shale or mudrock, silty gray shale and mudrock, silty shale, siltstone, and fine grained sandstone. This chapter briefly summarizes the history and correlation of the Genesee and related rocks.

Western and central New York were strongly glaciated during the Pleistocene by continental ice sheets. Although some of the topography was considerably modified by scour and erosion, particularly along main drainage lines in the vicinity of the Finger Lakes and Lake Erie, deposition generally was much greater than erosion. Thick sheets of till or glaciofluvial debris blanket much of the area and locally, as at Watkins Glen, filled the deeply incised valleys south of the lakes with as much as 1,200 feet of glacial sediment. Because of the extensive cover of glacial debris, the Devonian bedrock is rarely exposed, except in the cliffs along lake shores or along the steep-walled valleys of postglacial streams that have cut through the mantle of younger sediment. The scarcity of outcrops greatly hindered mapping in western New York, and the correlation of similar-appearing but nonequivalent facies led to much stratigraphic confusion in the past.

The paucity of outcrops and the subtle local variations in the gentle south regional dip necessitated detailed measurement of most long exposures. Detailed mapping was commonly required to bridge broad areas of glacial drift between well-exposed sections along the larger streams, particularly in the western part of the area between Lake Erie and the hilly tracts bordering the Genesee River and the smaller Finger Lakes. Outcrops are more abundant near the larger Finger Lakes in the eastern half of the area, and some relatively thin but extensive key beds, for example, the Bluff Point Siltstone Bed of the West River Shale Member of the Genesee, can be traced widely in many of the stream gullies near Keuka and Canandaigua Lakes. South and east of a line from Tully, Onondaga County, to Ithaca, Tompkins County, outcrops are relatively poor, except locally along the Tioughnioga River and its tributaries. Most of the high well-rounded hills in the southern part of the mapped area lack long continuous exposures of bedrock (**Figure 13**).

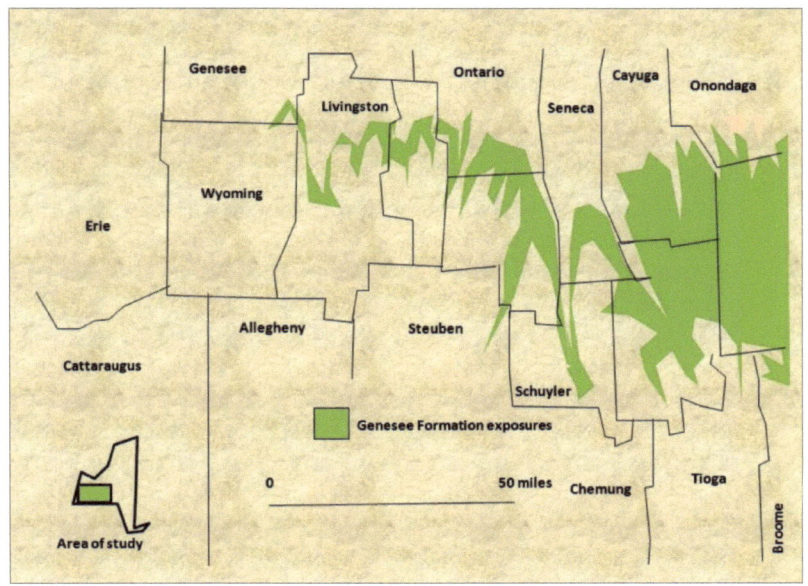

Figure 13. Location of exposures in Western and Central New York.

Where possible, surface data were supplemented with subsurface data from descriptions of drill cuttings obtained from scattered deep wells drilled for gas or oil. These subsurface data permitted compilation of structure maps and isopach maps of the Genesee Formation and some of its members. The combination of surface data, structure, and isopach maps greatly facilitated locating and tracing key beds and units in the Genesee Formation in the area east of Seneca Lake, where the Genesee contains a great thickness of silty shale and thick-bedded massive siltstone.

Rocks Below the Genesee Formation

From the place where it rises above the level of Lake Erie about a half mile west of the mouth of Pike Creek and eastward to the vicinity of Gage Gully, the Genesee Formation is underlain by the Moscow Shale, a sequence of medium-gray to medium-dark-gray fossiliferous calcareous mudrock. Ovoid, discoid, and irregularly ramose argillaceous limestone nodules are locally abundant in the Moscow. At places, the nodules coalesce to beds of impure limestone as much as 5 feet thick (Cazenovia Creek).

Small nodules and lenses of pyrite occur commonly in the upper part of the Moscow, and locally some beds of dark-grayish to brownish-black shale are intercalated in the formation. In the area between Pike Creek and Cazenovia Creek, a nodular bed of conodont-bearing argillaceous limestone as much as 0.15 foot thick, or North Evans Limestone, is found at the top of the Windom Member of the Moscow Shale; locally at Clover Bank it directly underlies the Genundewa Limestone Member of the Genesee Formation. Although these two carbonate units are locally in contact, their conodont faunas demonstrate a hiatus that is equivalent to as much as 125 feet of beds at Canandaigua Lake.

At many places between Pike Creek and Gage Gully, lenticular masses of abundantly fossiliferous pyrite as much as 0.6 foot thick and 1-10 feet long are present at the contact between the Moscow Shale and the Genesee Formation. Collectively, these lenses of pyrite are the Tully pyrite and the Leicester Marcasite Member of the Moscow Formation.

The age of the Leicester has been in question for a considerable time. Because in most outcrops, the lenses of pyrite are above the gray mudrock of the Moscow and at the base of the black shale or brownish-black mudrock of the Genesee, the Leicester could be assigned to either unit. At Gage Gully at the base of the Genesee Formation, nodules of pyrite of the Leicester-type were found to be half an inch thick and 0.5-1.5 feet long on nodules of Tully Limestone 0.8 foot thick and 2 feet long. The nodules of Tully Limestone, which contain *Hypothyridina venustula,* are the western feather edge of the main sheet of Tully. Clearly at Gage Gully, the Leicester is younger than Tully Limestone and the subjacent Moscow Shale. To the west, the age of the Leicester is equivocal because the Tully is absent, and the uppermost part of the Moscow Shale is progressively older westward across western New York.

The Moscow Shale, like many other units in the Devonian of New York, thins to the west and thickens to the east. The top of the Tichenor Limestone which marks the base of the Moscow Shale is exposed at water level at the mouth of Pike Creek and is 11 feet below the base of the Genesee Formation. At Elevenmile Creek, about 33 miles to the east, the Moscow has thickened to 55 feet. From the east side of Canandaigua Lake eastward to Tully, the Genesee Formation is underlain by the thick-bedded to massive cobble weathering fossiliferous Tully Limestone. The Tully is generally a slightly argillaceous, dark-gray to grayish-black, bluish-gray to tan weathering calcilutite with some interbedded calcarenite and wispy laminae of quartz silt. Locally at Borodino, the Tully contains abundant encrinal beds intercalated in a mound complex. Throughout much of its extent, the Tully contains many fossils, although their presence may be obscured by the cobbly weathering.

The resistant Tully generally forms the massive capstone of falls and cascades along its outcrop. Commonly, the formation is poorly exposed between streams, but locally on the hilltops near Tully and between Cayuga and Skaneateles Lakes, the limestone crops out sparsely in fields and ditches.

The Tully thickens from a zone of nodules about 0.8 foot thick at Gage Gully to about 25 feet in the vicinity of Tully. Subsurface data show that the formation thickens to the south and reaches a maximum of more than 200 feet in the vicinity of Lock Haven, Pa. The Tully is readily identified in well cuttings, and it makes a good marker unit for subsurface work. Subsurface correlations of the Genesee Formation in much of south-central New York are based upon the presence of the massive resistant Tully Limestone just below the black shale of the Geneseo Shale Member of the Genesee Formation.

The upper contact of the Tully is sharp and appears conformable from the eastern side of Canandaigua Lake to Cayuga Lake. The uppermost massive bed of the Tully is abruptly overlain by brownish-black shale of the Geneseo Shale Member. East of Cayuga Lake, the Tully is overlain by limestone and shale inter-fingering in a zone as much as 15 feet thick and consisting of brownish black slightly calcareous shale in 1- to 6-inch beds intercalated with argillaceous brownish-black to olive-black nodular limestone in beds 3-18 inches thick. Commonly, the argillaceous limestone does not resemble the subjacent cobbly weathering Tully. It weathers in irregular plates and slabs and is lithologically similar to the lenticular beds of coalesced limestone nodules that are present higher in the black or dark brown shale of the Genesee. The sequence of transitional beds in the Genesee Formation were included because they are lithologically unlike the Tully and are separated from the massive cobbly weathering Tully by brownish-black shale similar to the main body of the Geneseo Shale Member, the basal member of the Genesee Formation. These beds may be included in the Tully Limestone. On physical evidence, these transitional beds belong in the Genesee Formation rather than in the Tully Limestone.

Genesee Formation

The Genesee Formation was defined to include all the strata between the Tully Limestone below and the Middlesex Shale Member of the Sonyea Formation above in the area between Tully, Onondaga County, and Gage Gully on the east side of Canandaigua Lake where the Tully Limestone pinches out. West of Canandaigua Lake, the Genesee Formation is overlain by the Sonyea Formation and underlain by the Moscow Shale.

Prior to this redefinition, the strata included in our Genesee Formation had been included partly in the Genesee Shale, partly in the Portage Formation, and partly in the Genesee Group. The history of the usage of the name Genesee is relatively complicated and will not be discussed here.

In the area west of Canandaigua Lake, the Genesee is composed largely of black, grayish-black to olive-black shale and dark- to medium-gray shale or mudrock; in addition, the formation contains small amounts of nodular limestone, argillaceous siltstone, and an abundance of intercalated limestone nodules of many shapes and sizes. Four members are recognized. In ascending stratigraphic order, they are the Geneseo Shale Member, the Penn Yan Shale Member, the Genundewa Limestone Member, and the West River Shale Member. Tongues of siltstone and shaly siltstone wedge into the Genesee Formation between Canandaigua Lake and Cayuga Lake. The change from predominantly shale and mudrock to predominantly siltstone is well shown in sections along streams tributary to Seneca Lake between Dresden and Watkins Glen. In the vicinity of Ithaca at the south end of Cayuga Lake, two thick units of siltstone and an intercalated tongue of dark shale—the Sherburne Flagstone Member, the Renwick Shale Member, and the Ithaca Member—separate the Penn Yan Shale Member from the West River Shale Member.

The Genesee Formation is composed mainly of siltstone, shaly siltstone, and silty shale from Cayuga Lake to the east edge of the area studied. The Genesee Formation is about 9 feet thick where it plunges below the level of Lake Erie about half a mile west of the mouth of Pike Creek. The formation thickens eastward to about 130 feet near Geneseo, about 250 feet at Canandaigua Lake, and more than 940 feet near Ithaca. The Genesee thickens to the southeast in the subsurface and is more than 1,200 feet thick along the State line southeast of Elmira. As the Genesee Formation thickens to the south in the subsurface, the shaly units thicken. In north-central Pennsylvania the Genesee merges into and its identity is lost in a great mass of very dark gray to brownish-black slightly silty shale containing scattered beds of very dark gray to medium-gray siltstone.

The upper boundary of the Genesee Formation is placed at the base of the widespread grayish black to brownish-black shale of the Middlesex Shale Member of the Sonyea Formation. The contact is sharply marked and conformable throughout the area west of Ithaca, where the gray shale and intercalated thin-bedded siltstone in the upper part of the West River Shale Member of the Genesee contrast markedly with the overlying black shale in the Middlesex Shale Member of the Sonyea.

East of Ithaca, in parts of Cortland, Tioga, and Broome Counties, the Middlesex is relatively thin and rests upon thick-bedded siltstone and shaly siltstone in the upper part of the Ithaca Member, which mark the top of the Genesee Formation there. The boundary between the Genesee and the Sonyea Formations is clearly discernible at the change from siltstone to dark-brown silty shale but good exposures of the stratigraphic interval containing these beds are scarce.

Geneseo Shale Member

Geneseo Shale consists of 83 feet of "black" shale separating the Moscow Shale from the Genundewa Limestone at Fall Brook near Geneseo. Much of the 83 feet of strata was dark gray massive mudrock and that only two beds of brownish-black shale, totaling 10.5 feet in thickness, might be traced east from Fall Brook into the thick black shale in the base of the Genesee Formation in the Finger Lakes district. Consequently, the name Geneseo is restricted to the two tongues of black shale in the lower part of the 83-foot sequence of the Geneseo at Fall Brook and selected the 44 feet of black shale overlying the Moscow Shale at Menteth Gully as the reference section for the Geneseo Shale Member of our Genesee Formation.

The upper of the two beds of brownish-black shale at Fall Brook is not a tongue from the main mass of the black Geneseo Shale Member at Menteth Gully. Rather, it is a discrete bed of black shale in the Penn Yan Shale Member. The Geneseo Shale Member Is restricted to the basal 5 feet of brownish-black shale in the Genesee Formation at Fall Brook and include the upper bed of brownish-black shale in the Penn Yan Shale Member.

The Geneseo is composed predominantly of grayish-black, brownish-black, and olive-black fissile shale. The rock is laminated and massive on fresh exposure and becomes fissile upon weathering. Commonly, it weathers to small sharp-edged chips or plates stained reddish brown or orange by limonite derived by weathering of contained pyrite. Locally, beds of very dark gray or brownish-gray shale are intercalated in the Geneseo mainly in the upper part of the member in the western part of the mapped area. Grayish-black slightly argillaceous limestone nodules, which weather medium gray to medium light gray, are locally abundant in zones in the Geneseo. The nodules range in shape from spheroidal to discoidal and in diameter from half an inch to 3 feet. Many of the larger nodules show a septarian structure of white calcite and small amounts of barite, dolomite, ankerite, siderite, pyrite, and marcasite. Locally, the nodules coalesce to form lumpy-surfaced beds of argillaceous limestone 1-18 inches thick.

Medium-gray laminar to wavy-bedded siltstone in layers a quarter of an inch to 10 inches thick occurs sparingly in the Geneseo. Some beds of siltstone are calcareous and may contain scattered calcareous nodules, whereas other layers are noncalcareous and strongly resistant to erosion. The number of siltstone beds in the Geneseo increases eastward from Keuka Lake to the eastern border of the mapped area.

The Geneseo Shale Member forms the basal part of the Genesee Formation from the Tully quadrangle westward across the study area to the Depew quadrangle in eastern Erie County. The massive black shale is about 80 feet thick near Tully. It reaches its maximum thickness of about 130 feet in surface exposures at Glenwood Creek and Taughannock Creek northwest of Ithaca, and it thins westward to 44 feet at Menteth Gully. The Geneseo Shale Member is present at the base of the Genesee Formation as far west as Cayuga Creek, and it feathers out into dark-gray shale and mudrock of the Penn Yan Shale Member between Cayuga and Buffalo Creeks. The Geneseo Shale Member thickens rapidly from the east border of the mapped area to a maximum in parts of Chemung, Tompkins, Schuyler, and Steuben Counties. The member thins more gradually to the west and northwest and is absent in the western part of the mapped area.

Fossils are relatively scarce in much of the Geneseo Shale Member. However, in the area from Bellona eastward across the Ovid quadrangle and into the western part of the Genoa quadrangle to the Hubbard quarry adjacent to Lively Run, a small fauna of thin-shelled forms dominated by the inarticulate brachiopod *Orbiculoidea lodiensis* is found in the uppermost beds of the black shale of the Geneseo and in the basal dark-gray beds of the overlying Penn Yan Shale Member. The interval containing *Orbiculoidea lodiensis* is thickest in the vicinity of Lodi and Ovid; it thins to the east and west. *Orbiculoidea* was not located on the east shore of Cayuga Lake and in outcrops along the valley of Salmon Creek.

The upper boundary of the Geneseo Shale Member is generally well defined. The change from brownish-black shale of the Geneseo into dark-gray or dark-brownish-gray slightly silty mudrock and shale of the Penn Yan Shale Member is abrupt at many places and can be readily noted, particularly in cliff sections where weathering accentuates the difference in types of rocks. Subsurface data show that south of Tully and east of Ithaca, the Penn Yan Shale Member grades laterally into the Sherburne Flagstone Member, and resistant thick-bedded siltstone of the Sherburne directly overlies the Geneseo Shale Member. The boundary between the Geneseo and Sherburne is marked by the change from silty brownish-black shale to medium-gray siltstone.

The name Penn Yan was given to a 150-foot sequence of dark-gray mudrock and shale containing beds of black shale, calcareous nodules, and light-gray siltstone that overlies the Geneseo Shale Member and underlies the Crosby Sandstone in stream gullies along the south side of Keuka Lake outlet in the vicinity of Seneca Mills. The Penn Yan beds were believed to be a tongue of the West River Shale. This study, however, showed that the Penn Yan rocks are separated from the West River Member of the Genesee Formation by the Genundewa Limestone Member, the Crosby Sandstone, or by a thick wedge of clastic rocks, which include the eastern equivalent of the Genundewa. Consequently, the Penn Yan was designated as a member of the Genesee Formation.

The Penn Yan Shale Member consists mainly of dark- to light-gray, olive-gray to light olive-gray, and some greenish gray slightly silty mudrock and shale. Scattered beds of brownish-black iron stained shale and a profusion of spheroidal to discoidal argillaceous grayish-black limestone nodules are interbedded in the Penn Yan. Scattered laminae and beds of siltstone 0.05 to 4 feet thick range from nonresistant calcareous and argillaceous layers to resistant durable beds intercalated in the member. Siltstones are fewest in the western part of the mapped area and increase in number and thickness eastward to the vicinity of Tully. Most of the beds of siltstone are relatively lenticular and of local extent, but some of the more massive beds are sufficiently extensive for short range correlation.

The Penn Yan Shale Member overlies the Geneseo Shale Member throughout the outcrop belt from Bucktail Falls near Tully westward to Cayuga Creek, where the Geneseo is overlapped. Westward to Lake Erie the Penn Yan facies rests upon the Windom Member of the Moscow Shale or on the North Evans Limestone. The Penn Yan is present below the Genundewa Limestone Member west to the vicinity of Buffalo Creek but is absent locally in the area between Buffalo Creek and Eighteenmile Creek. In cliffs at the mouth of Pike Creek, a tongue of Penn Yan less than 1 foot thick separates the North Evans Limestone from the Genundewa Limestone Member of the Genesee Formation. The Penn Yan thickens from less than 1 foot at Lake Erie to a maximum in outcrop of about 155 feet at the type section at Keuka Lake outlet. East and southeast of the type exposure, the Penn Yan thins to 100 feet or less as the upper half of the member intertongues with and grades into the western part of a thick mass of siltstone, the Ithaca Member of the Genesee Formation.

East of Seneca Lake, the part of the Penn Yan equivalent to the lower half of the member at Keuka Lake outlet intertongues with the Sherburne Flagstone Member, and the Penn Yan pinches out of the section as a recognizable unit in the Tully quadrangle. Subsurface data show that the Penn Yan Shale Member thickens south of the Finger Lakes and is more than 200 feet thick below Van Etten in northeastern Chemung County. In the subsurface, the member is thin in western New York and cannot be identified with certainty in western Cattaraugus and Chautauqua Counties.

Thick-bedded calcareous or noncalcareous siltstone intercalated in the shaly Penn Yan at several localities has led to some vexing and erroneous correlations in the past. Several 2-foot-thick beds of argillaceous, calcareous, poorly bedded siltstone or very silty argillaceous limestone containing scattered spheroidal nodules of limestone and abundant *Styliolina fissurella* occur in the upper few feet of the Geneseo Shale Member and in the basal 10 feet of the Penn Yan Shale Member of Fir Tree Point on the west side of Seneca Lake. These siltstones were identified as "Genundewa" because of their lithologic and paleontological similarity to the Genundewa Limestone Member at Canandaigua Lake.

"Genundewa" was identified as somewhat similar-appearing bed of silty argillaceous limestone that caps the Geneseo Shale Member about 6.5 miles north, in the vicinity of Baskins Point. Because both localities are well east of the eastern edge of the sheet of Genundewa, it was assumed that "Genundewa" overlay the Genesee Shale, the unit was later defined as the Geneseo Shale Member, at Seneca Lake. At Baskins Point and Fir Tree Point, the "Genundewa" beds are about 180-220 feet below the Crosby Sandstone which is the lateral equivalent of the Genundewa Limestone Member at Seneca and Cayuga Lakes. This is not a part of the Genundewa Limestone of central and western New York. As a consequence of this misidentification of the Genundewa at Seneca Lake, the Sherburne and Ithaca were thought to be younger than the Genundewa, but in fact the Sherburne and at least the lower third of the Ithaca at Seneca and Cayuga Lakes are older than the Genundewa.

Several beds of coalesced argillaceous limestone nodules and two or three thick-bedded layers of siltstone are found in the basal part of the Penn Yan and cap the Geneseo Shale Member in the area from Bellona eastward across Seneca and Cayuga Lakes into the valley of Salmon Creek near Lansingville. Some of these beds are within the interval containing *Orbiculoidea lodiensis* near Lodi, Ovid, and Hubbard's quarry. Outcrop data show that these resistant beds are lenticular and that individual beds have only local extent. They appear to be oldest in the eastern part of the area and youngest near Bellona at the west.

A bed of silty argillaceous limestone nodules intercalated in calcareous argillaceous siltstone in Mill Creek at Lodi was observed at the top of the Geneseo Shale Member and described the fauna from it. The fauna became known as the *Orbiculoidea lodiensis* fauna, and the limestone nodules were called the Lodi Limestone. The "Lodi Limestone" was correlated with beds identified as "Genundewa" at Fir Tree Point because of faunal similarities and because both were at the top of the black-shale sequence. This correlation implied limestone at Lodi was equivalent to the Genundewa at Canandaigua Lake, whereas sections show that the nodular limestone in Mill Creek at Lodi is at least 195 feet below the horizon of the Genundewa, which at Lodi is represented by the sparsely fossiliferous Crosby Sandstone. Because several well-known faunal zones, such as the *Reticularia laevis* zone near Ithaca, occur in the stratigraphic intervals between beds containing the *Orbiculoidea lodiensis* fauna and the Genundewa-Crosby bed, paleontological studies in which the Genundewa fauna has been equated with the *Orbiculoidea, lodiensis* fauna should be reevaluated.

The upper boundary of the Penn Yan Shale Member from Buffalo Creek, Erie County east to the vicinity of Shuman Cemetery, Ontario County, and Corry Gully, western Yates County, is at the base of the thin sheet of the styliolinid Genundewa Limestone Member. The top of the Penn Yan probably closely approximates a time line across this part of western New York. From the western side of Keuka Lake eastward, the top of the Penn Yan is placed at the base of an eastward-thickening sequence of siltstone, the combined Ithaca, Renwick Shale, and Sherburne Flagstone Members. A lower tongue of Penn Yan extends far to the east and interfingers with the Sherburne Flagstone Member in the vicinity of Tully, Onondaga County. The middle and upper parts of the Penn Yan interfinger with the Sherburne Flagstone, Renwick Shale, and Ithaca Members between Keuka and Cayuga Lakes. Consequently, the top of the Penn Yan is placed at progressively older stratigraphic positions to the east. In contrast to the close approach to a time line west of Keuka Lake, the top of the Penn Yan is sharply diachronic to the east.

Sherburne Flagstone Member

The name Sherburne was proposed for the sequence of silty and sandy rocks above the black shale that overlies the Tully Limestone in the vicinity of Sherburne, Chenango County. The Sherburne was excluded from the Portage Formation, and assigned to a member status in the Genesee Formation. From Tully west to the vicinity of Ithaca and Cayuga Lake, the Sherburne is composed largely of thin-bedded to massive laminar light-gray siltstone, and some silty shale, shaly siltstone, and a few beds of very fine grained sandstone.

Commonly, the bedding ranges from an inch to 1 foot in thickness, but locally some massive beds exceed 15 feet. Many of the thicker beds are structureless or slightly laminated in the basal part and ripple marked near the top. Commonly, the base of the siltstone bed is sharp, whereas the top is gradational into the thin shaly parting above. Sole markings are conspicuous on the base of many beds, and small-scale cut-and-fill structures are locally abundant. The Sherburne shows depositional features characteristic of a turbidite accumulation. Sequences of thick or thin beds appear to have some lateral continuity, but individual beds generally cannot be correlated in adjacent sections with certainty.

The Sherburne Flagstone Member is presently overlying the Penn Yan Shale Member from the Tully quadrangle west to the ridge between Cayuga and Seneca Lakes where it intertongues with and grades laterally into the silty mudrock and shale of the Penn Yan Shale Member. The Sherburne is about 155 feet thick near Kelloggsville, 110 feet thick at Glenwood Creek northwest of Ithaca, where the member is designated as a reference section for the member, and about 45 feet thick at Sheldrake Creek. The Sherburne is not a recognizable unit at Mill Creek near Lodi.

About 40 feet of the Sherburne is present in a drill core cut near Corbett Point north of Watkins Glen on the west side of Seneca Lake, and some of the top of the member is exposed below a 2- to 3-foot massive siltstone, horizon "S" in Quartermile Creek and Excelsior Glen at the south end of Seneca Lake. Subsurface data are scant, but the isopach of the Sherburne Flagstone Member shows the member to be more than 250 feet thick in parts of Cortland and Tioga Counties. The abrupt thinning of the wedge of Sherburne in the vicinity of Cayuga Lake marks the zone where the distal edge of the Sherburne turbidite fan interfingers with the Penn Yan. The Sherburne can be identified with certainty in the subsurface only where the overlying Renwick Shale Member is definitely present. Elsewhere, it cannot be separated from the overlying Ithaca Member. Fossils are not abundant in the Sherburne in most of its extent in the study area, although locally they may be found in moderate numbers. The first *Reticularia laevis* zone is found in the upper 35-40 feet of the member at Renwick Brook and at Willow Point Gully.

The Sherburne Flagstone Member is overlain by the Renwick Shale Member of the Genesee Formation from Threemile Point Gully west across the area. The upper boundary of the Sherburne is the base of the oldest bed of brownish black or very dark olive gray shale in the Renwick Shale Member. Locally, as at Glenwood Creek and Buck Corners Creek, the contact is clearly defined; elsewhere, as at Sheldrake Creek, the top of the Sherburne is difficult to identify.

Renwick Shale Member

At Cayuga Lake, the Sherburne Flagstone Member is overlain by a 5- to 45-foot sequence of grayish-black, brownish-black to very dark olive-gray iron-stained shale containing an abundance of siltstone-filled scour channels. Near Ithaca, this sequence of dark shaly strata was informally named Ithaca *Lingula* shale and was apparently designated as the Renwick Shale Member of the Middlesex Shale Formation. Because the Renwick was proposed in an abstract, the type section and boundaries of the unit were not described. Later studies clearly showed that the Renwick was not a part of the Middlesex Shale Member of the Sonyea Formation. The Renwick was redefined as a member of the Genesee Formation and selected outcrops in Renwick Brook north of Ithaca as the type exposure of the member.

The Renwick Shale Member is present at Threemile Point Gully and Scott Gulf south of Skaneateles Lake, where it consists largely of very dark brown to dark-olive-gray iron-stained silty shale containing a scattering of scour channels filled with medium- to light-gray siltstone. Individual channels may be as much as 3 feet deep and 30 feet wide. Channel filling may pinch out at the edge of the channel or merge into a sheet of laminated or ripple bedded siltstone. In much of the area between Skaneateles and Owasco Lakes, the dark shale in the Renwick is not as conspicuous as in the member west of Cayuga Lake, and the Renwick can best be identified by the abundance of siltstone-filled scour channels.

The Renwick is about 45 feet thick at Threemile Point Gully and thickens to more than 60 feet near Ithaca. In the vicinity of Ithaca, the amount of brownish-black and very dark gray rock increases considerably, and the number of scour channels reaches a maximum. Scant data indicate that the greater number of channels trend southwest and have current flow from east to west, although some channels show current flow to the east.

Between Cayuga and Seneca Lakes, the Renwick Shale Member thins relatively abruptly, and the amount of intercalated siltstone also decreases. The Renwick is composed mainly of very dark gray and brownish-black shale containing a few laminae or stringers of siltstone on the west side of Seneca Lake where the member is generally less than 10 feet thick. The isopach of the Renwick Shale Member shows a linear maximum of more than 60 feet in western Tompkins and eastern Chemung Counties. The member thins rather rapidly to the northwest, except for a local protuberance from the area of greatest thickness across part of western Tompkins County to the southern part of Seneca Lake.

The westward-thinning Renwick was traced into the type area of the Penn Yan Shale Member at Keuka Lake outlet and to Sartwell Ravine south of Penn Yan where the Renwick consists of 6-8 feet of brownish-black iron-stained fissile shale in the middle of the Penn Yan. Most probably the Renwick is equivalent to one of the thinner beds of black shale in the Penn Yan farther west, but we were unable to establish a positive correlation west of Keuka Lake.

At Cayuga Lake, the Renwick Shale Member is sandwiched between massive siltstone units, the Sherburne Flagstone Member below and the Ithaca Member above. Between Cayuga and Seneca Lake, the Sherburne grades laterally and interfingers into the shaly Penn Yan and the base of the Ithaca Member rises stratigraphically to the northwest. At Quartermile Creek at Watkins Glen, siltstone and silt shale of the Ithaca cap the Renwick, whereas near Lodi, 15 miles to the north, about 25 feet of Penn Yan separates the Renwick from the younger Ithaca. Along Keuka Lake outlet, as much as 75 feet of Penn Yan shaly strata separates the Renwick from the Ithaca.

A second extensive bed of dark-brownish-gray iron-stained shale is present stratigraphically higher in the Penn Yan Shale and Ithaca Members in the vicinity of Seneca Lake. This bed of dark shale, here informally named the "Starkey" black bed "Stb," is typically exposed in the stream east of Starkey where the bed is about 9 feet thick. Recognition of the presence of two units of iron-stained dark-brownish-gray to grayish-black shale in the upper half of the Penn Yan and equivalent beds around Seneca Lake is essential to working out the local stratigraphy, particularly along the west side of Seneca Lake, where locally the Renwick dips below lake level near Glenora Falls and in the vicinity of Salt Point.

Throughout much of its extent, the Renwick Shale Member is unfossiliferous. However, at Renwick Brook and in adjacent sections, the Renwick contains *Lingida, Leiorhynchus,* and *Pluminaria plumaria* in considerable numbers. The upper boundary of the Renwick is much less distinct than the lower throughout most of the area east of Seneca Lake. Commonly, the upper boundary is placed at the top of the youngest siltstone filled channel in the sequence of interbedded dark shale and intercalated siltstone. Obviously, this boundary is not at the same stratigraphic position throughout the area. West of Seneca Lake, where the Renwick is within the Penn Yan, both top and base are sharply defined by the change in color from grayish or brownish black of the Renwick to the medium olive gray of the Penn Yan.

The Renwick Shale Member is overlain eastward from Seneca Lake by an eastward-thickening wedge of siltstone, silty shale, and silty mudrock. This mass of silty rock makes up the Ithaca Group and much of the Ithaca Shale Member of the Portage Formation. The name Ithaca is restricted to the silty rocks between the Renwick Shale Member below and the West River Shale Member above in the area of Cayuga Lake. Because much of the member is non-shaly rock in this area, the descriptive term "shale" was deleted from the member name. The Ithaca Member is composed largely of medium- to light-gray, tan-weathering, slightly argillaceous medium- to coarse some very fine to fine-grained quartz sandstone. Medium-gray to greenish-gray silt-shale, shaly siltstone, silty shale, and silty mudrock are found in subordinate amounts. A few beds of dark-gray to brownish-black commonly iron-stained shale and scattered beds of silty coquinoidal limestone are intercalated in the Ithaca.

Most of the siltstones in the Ithaca Member are turbidites. Bedding ranges in thickness from laminae less than 0.1 inch to more than 10 feet. Many beds have a discrete base and an irregular or indefinite top. They show basal-graded, lower laminated, and current-rippled divisions similar to those of the turbidites in the Sonyea Formation. Many of the thicker beds show both parallel laminations and complex cross laminations, whereas the thinner layers may be largely ripple bedded. Sole markings are abundant, and small-scale current scours and cut-and-fill structures are common mainly in the eastern part of the area. Some of the more massively bedded siltstone contains sub-spheroidal to lenticular calcareous nodules as much as 2 feet thick and 4 feet long.

Locally, lenses of silty coquinoidal limestone or lenses of broken fossils are intercalated in the member. Commonly, sequences as much as 25 feet thick of thick-bedded to massive siltstone are intercalated in somewhat thicker zones of thin-bedded siltstone, silt-shale, and silty mudrock. Although individual beds may have considerable lateral extent, the individual zones of massively bedded siltstone are generally lenticular and do not have regional continuity. An exception to this statement is the Crosby Sandstone, whose regional continuity was later demonstrated.

The Ithaca Member is about 450 feet thick in the vicinity of Ithaca, Willow Point, and only about 30 feet thinner at Watkins Glen at the south end of Seneca Lake. The member thins to the northwest and is about 275 feet thick at Lodi on the east side of Seneca Lake and about 200 feet thick at Plum Point Creek near Himrod. The Ithaca is about 100 feet thick at Penn Yan.

The member is recognizable as far west as Chidsey Point Gully and intertongues and grades laterally into the West River Shale Member between Penn Yan and West River valley. The Ithaca is not a recognizable unit in sections at Voak and Shuman Cemetery.

The Ithaca Member thickens east of Ithaca, but outcrop data are scant in this part of the study area. The great width of outcrop belt of the Genesee Formation, the lack of long continuous exposures, and the absence of identifiable key beds in the Ithaca east of Tompkins County prevent an accurate determination of the thickness of the Ithaca in this area.

The isopach map of the Ithaca Member shows thickness of more than 450 feet in eastern Chemung and western Tioga Counties. Subsurface data are virtually absent to the east in eastern Tioga, Broome, and Cortland Counties, thus preventing an accurate determination of the thickness of the member in the southeastern corner of the study area. In the subsurface, the Ithaca thins to the northwest from Ithaca and Watkins Glen to a zero line some place in central Yates County, northwest of Penn Yan. The general configuration of the wedge of Ithaca rocks suggests deposition from a southeastern source in contrast to a more easterly source for the older Sherburne Flagstone Member.

In contrast to many of the more massive siltstones in the Ithaca which are lenticular and of local extent, the Crosby Sandstone, a 3- to 10-foot thick massive unit of coarse-grained siltstone and some very fine grained sandstone, is a key unit of great areal extent. This unit was traced from Penn Yan east and southeast to Watkins Glen and to the vicinity of Ithaca. It forms the basal beds of the Ithaca Member along Keuka Lake outlet east of Penn Yan and along the east arm of Keuka Lake south of Penn Yan to Sunset Point Gully. Originally, the Crosby was described as being exposed near the community of Crosby south of Penn Yan on the east side of Keuka Lake. Consequently, the Crosby is not exposed in Crosby Gully, because the south dip places the unit below lake level. Although their Crosby is not exposed at Crosby Gully, it is a strongly erosion-resistant unit that is exposed at many places south and east of Penn Yan.

The Genundewa conodont fauna was identified in the Crosby near Penn Yan and later at other localities to the east and south as its stratigraphic position was projected towards Watkins Glen and Ithaca. Ultimately, the Crosby was established as a datum from Penn Yan to Ithaca through the thick and relatively indivisible sequence of silty rocks of the Ithaca Member.

Although the Ithaca is relatively unfossiliferous, locally some beds contain moderate numbers of fossils, particularly in the eastern part of the area. One of the more spectacular accumulations of fossiliferous material is the Williams Brook coquinite member, a 10-foot-thick interval of shell fragments intercalated in the sequence of thick-bedded to massively bedded siltstone. The Williams Brook coquinite is about 70-80 feet below Torrey's Crosby Sandstone and about 165 feet above the base of the Renwick Shale Member at Williams Brook. At this site beds that consist of cross-laminated shell fragments and that are as much as 3 feet thick form a small waterfall. Similar accumulations of fossils were not located at the same horizon in adjacent sections.

In the lower part of the Ithaca Member at Big Stream, a bed of calcareous siltstone containing many small fossils was identified as the Parrish Limestone Lentil of Cashaqua Shale. This bed is a local fossiliferous siltstone, which is about 30 feet above the Crosby Sandstone. The true Parrish Limestone Lentil of the Cashaqua crops out within the Cashaqua Shale Member of the Sonyea Formation farther west in Big Stream at a much higher elevation. In the area east of Seneca Lake, where the Ithaca Member overlies the Renwick Shale Member, the lower boundary of the Ithaca is arbitrarily placed at the top of the youngest siltstone-filled channel in the sequence of dark shale and intercalated siltstone that composes the Renwick. West of Seneca Lake, the Ithaca overlies the Penn Yan Shale Member. The base of the Ithaca is marked by thick bedded to massive siltstone or the Crosby Sandstone on a sequence of medium-gray mudrock containing subordinate amounts of black shale and many calcareous nodules.

The upper boundary of the Ithaca is arbitrarily placed at the top of the youngest sequence of thick bedded Ithaca-type siltstone above which the sequence is largely composed of shale and mudrock of the West River Shale Member. The top of the Ithaca is diachronous and becomes younger to the south and southeast from Penn Yan to Watkins Glen and Ithaca. Locally along Seneca Lake north of Watkins Glen, a tongue of massive Ithaca-type siltstone is found in the upper part of the West River Shale Member not far below the base of the Middlesex Shale Member, the basal member of the Sonyea Formation. This tongue of siltstone is horizon "MF" which was originally assigned to a position above the Middlesex rather than below the member.

East of Ithaca, across the Dryden and Harford quadrangles, data are scant. The top of the Ithaca appears to rise stratigraphically in the section, although the West River Shale Member is thick in the central and eastern parts of the Dryden quadrangle.

Many beds of siltstone occur in the West River Shale Member in this area, and differentiation of the Ithaca from the West River becomes increasingly difficult to the east. In eastern Tioga and western Broome Counties in the Harford quadrangle, Ithaca-type siltstones are found not far below the basal brownish-black shale of the Middlesex Shale Member. Although the data are scant for want of long continuously exposed sections in the upper part of the Genesee Formation, the Ithaca Member appears to be the upper member of the Genesee Formation along the east side of the Tioughnioga River valley. The West River Shale Member appears to have graded laterally and intertongued with the Ithaca Member along the eastern border of the study area.

Genundewa Limestone Member

The Genundewa Limestone Member of the Genesee Formation was originally named the Genundewa Limestone and later was redesignated the Genundewa Limestone Lentil of the Geneseo Shale. The Genundewa was not a lentil in the Geneseo Shale, and was changed to member status in the Genesee Formation, although, because it is extremely thin in comparison with other members of the Genesee, it cannot be mapped at a scale of 1:24,000. The Genundewa is typically exposed at Genundewa Point at the southwest end of Bare Hill, known locally as "Genundewah" or Genundewa Hill, on the east side of Canandaigua Lake.

As originally defined and as redefined, the Genundewa at its type locality consists of a 12 to 15-foot-thick sequence of dark-gray calcareous shale and mudrock containing many nodules of very dark gray to black argillaceous limestone from 1 to 10 inches thick, several beds of grayish-black to brownish-black slightly silty shale, and several zones of spheroidal to discoidal limestone nodules as much as 8 inches thick and 1.5 feet in diameter. The uppermost bed of limestone is composed largely of the shells of the minute pteropodlike *Styliolina fissurella* set in a matrix of calcilutite. *Styliolina* is much less abundant in the calcareous beds and nodules below.

The Genundewa is further restricted to only the relatively thin bed of styliolinid limestone at the top of the 12- to 15- foot-thick sequence heretofore identified as Genundewa at Genundewa Point. The remainder and much larger part of the sequence are hereby included in the Penn Yan Shale Member of the Genesee Formation. Detailed sections and microcorrelations indicate that the sheet of styliolinid Genundewa extends as a single unit ranging in thickness from 0.1 to 2 feet from the shore of Lake Erie one-half mile west of the mouth of Pike Creek east to the vicinity of Corry Gully and Shuman Cemetery, a distance of about 95 miles.

Throughout its extent, the Genundewa consists of one or more layers of grayish-black to dark-olive-gray slightly argillaceous, slightly silty, dense styliolinid limestone in which shells of *Styliolina fissurella* may make up as much as 70 percent of the rock. Locally, some layers contain small pellets or pseudo-oolites. Freshly broken surfaces of the Genundewa have a speckled glistening surface produced by the calcite filling of the randomly oriented pteropod shells. The Genundewa generally weathers to sheets, nodules, and slabby cobbles of light-gray or tan-surfaced limestone.

Bedding is irregular, undulose, or nodular. In general, the basal surface of the member is more irregular than the top. At many places, large discoidal nodules as much as 0.6 foot thick and 1-2 feet in diameter occur in the upper part of the Penn Yan just below the Genundewa, and the Genundewa bed is slightly arched over the centers of the nodules. Locally, many small spheroidal or discoidal nodules are found in the Genundewa, and at many places, particularly east of Canandaigua Lake, lenticular laminae and stringers of quartz silt as much as 0.05 inch are present in the member. Although rarely more than 1 foot thick, the Genundewa has a characteristic lithology that makes it an easily identified key bed. Because of its superior resistance to erosion in comparison with the soft enclosing shale and mudrock, the Genundewa caps small falls and riffles in tributary creeks and locally caps some spectacular falls.

The Genundewa Limestone Member can be traced readily from Lake Erie east to Corry Gully in the West River valley east of Canandaigua Lake. Although data are scant between Corry Gully and the gully at Penn Yan, the solid sheet of Genundewa at Corry Gully appears to thin and to grade laterally into a digitate sequence of lenses of silty argillaceous styliolinid limestone intercalated with some calcareous silty shale, with siltstone laminae, and a scattering of medium to large discoidal silty nodules in the vicinity of Shuman Cemetery. Exposures of the Genundewa interval are scarce between Shuman Cemetery and Penn Yan, but physical and paleontological data show that the Genundewa Limestone Member is essentially the stratigraphic equivalent of the Crosby Sandstone at Penn Yan. Actually, the featheredge of their Crosby may be the thin siltstone capping the Shuman Cemetery and the thin siltstone laminae in the upper part of the Genundewa at Corry Gully.

Correlation of the Genundewa Limestone Member with the Crosby provides a datum between Lake Erie and Ithaca that crosses several facies and divides the Genesee Formation vertically into about equal parts. The Genundewa Limestone Member is probably the most fossiliferous member in the Genesee Formation, in spite of its relatively slight thickness.

In addition to the numerous styliolinids, the Genundewa contains many conodonts, a fairly large fauna of invertebrate macrofossils, some fish bones and teeth, and a scattering of plant fossils. The sea at the time the Genundewa accumulated must been swarming with life, at least in the upper part of the water column, and the adjacent land area was covered in part with primitive terrestrial plants. Commonly, the upper boundary of the Genundewa is sharply defined by the abrupt change from styliolinid limestone to dark-gray slightly calcareous shale containing some siltstone laminae. At many places, the basal bed of the West River Shale Member is a slightly calcareous dark-gray argillaceous siltstone as much as half an inch thick. At other localities, medium-gray to very dark olive-gray shale lies directly upon the slightly undulose upper surface of the Genundewa Limestone Member.

West River Shale Member

The name West River Shale was applied to a sequence of shaly rocks above their Genundewa limestone and under their Standish flags and shales in West River valley about 3 miles east of Canandaigua Lake. Standish flags and shales are not lithologically different from much of the West River strata below. The West River was redesignated as a member of the Genesee Formation and included the Standish in the redefined West River Shale Member. The West River Shale Member is composed largely of dark-gray to medium-light-gray shale and mudrock. The strata become more silty eastward, and thin-bedded light-gray shaly siltstone is common in the member east of Ithaca. Many relatively thin beds of grayish-black to dark-olive-gray iron-stained shale an inch to several feet thick are intercalated in the member, especially in the area west of Seneca Lake. Because these individual black beds have great lateral continuity, they are excellent key beds for widespread correlation.

Many layers of spheroidal to discoidal grayish-black argillaceous, slightly silty limestone nodules 1 inch to 2 feet thick and 6 inches to 4 feet in diameter are present in the West River. Commonly, the layers or zones of nodules are associated with beds of black shale and are useful for correlation of adjacent sections. Several layers of grayish-black bituminous to medium light-gray slightly calcareous siltstone are intercalated in the member west of Keuka Lake. Siltstones become thicker bedded and more numerous in the West River Shale Member east of Keuka Lake, although they do not become a major constituent of the member in the area west of Ithaca.

The West River Shale Member is about 8 feet thick in the vicinity of the mouth of Pike Creek at Lake Erie. The member thickens eastward to about 53 feet on the west side of the Genesee valley and is about 132 feet thick at Seneca Point Gully, the reference section for the member. The West River thins to about 95 to 100 feet at Keuka Lake and is about 75 feet thick at Watkins Glen. It thickens to more than 130 feet near Ithaca and may be as much as 200 feet thick in the north-central part of the Dryden quadrangle. The West River appears to thin to the east across the Harford quadrangle by intertonguing with the Ithaca Member and is less than 50 feet thick along the west side of the Tioughnioga Valley. The isopach of the West River Shale Member shows an eastward thickening wedge of rock with a maximum thickness of more than 200 feet in the vicinity of Van Etten in northeastern Chemung County. The area of thick West River appears to extend northeast to the thick sections in the north-central part of the Dryden quadrangle. The abrupt thinning of the West River across the Harford quadrangle cannot be demonstrated in the subsurface because of the lack of control.

The West River Shale Member contains some thin beds that have great lateral continuity. Of these beds, the Bluff Point Siltstone Bed is the most widespread. The Bluff Point Siltstone Bed, also known as the Keuka Flagstone and the Bluff Point Flagstone Member of the Standish is a resistant, medium- to light-gray convolute-bedded siltstone 3-4 inches thick, which has a strongly ripple marked upper surface in outcrops around Keuka Lake. Because it is resistant to erosion, the Bluff Point Siltstone Bed commonly caps small falls and forms a riffle in broad creek beds. It is a fine key bed for local mapping. The Bluff Point was traced east to Plum Point Creek west of Himrod and tentatively identified it on Mill Creek at Lodi. The Bluff Point is present south along the west side of Seneca Lake to Big Stream-Glenora and appears to merge into the upper part of the Ithaca between Big Stream and the creek at Reading Center.

The Bluff Point loses its convolute bedding between Keuka Lake and Canandaigua Lake, where it is represented by a much cross laminated quartzitic dark-gray siltstone about 1.5 to 2 inches thick. Correlation of the many beds of black shale and nodules in the West River Shale Member confirms the extension of the Bluff Point between Keuka and Canandaigua Lakes. The thin, resistant highly ripple marked dark-gray Bluff Point Siltstone Bed is present in most sections between Seneca Point Gully at Canandaigua Lake and Eighteenmile Creek southwest of Buffalo. The absence of the Bluff Point in the Honeoye quadrangle may be the result of non-deposition over a local high spot in the sea floor when the Bluff Point silt flushed across the sea bottom. The presence of this unique turbidite in the middle part of the West River Shale Member gives us a time line across western and central New York from Lake Erie to Seneca Lake.

The Bluff Point Siltstone Bed is only slightly less extensive than the Genundewa-Crosby horizon.

The Standish flags and shales consist of three beds of dark gray to dark-olive-gray bituminous siltstone or very silty brownish-black shale about 1-3 inches thick and some associated limestone nodules, which occur in the upper 4 feet (Buck Run Creek) to 7 feet (Chidsey Point Gully) of the West River Shale Member between the Genesee River and Keuka Lake. The three beds can be correlated fairly well between adjacent sections, but the siltstone and enclosing shale are not sufficiently different lithologically to warrant recognition as a discrete member. Other beds of siltstone or black shale in the West River has comparable areal extent.

The upper boundary of the West River Shale Member throughout most of the mapped area is the base of the brownish-black to olive black fissile shale of the Middlesex Shale Member of the Sonyea Formation. The contact between the West River and the Middlesex is sharp throughout the mapped area. In the extreme eastern part of the mapped area, where the east-thinning wedge of West River Shale Member intertongues with the Ithaca Member, the upper boundary of the Genesee Formation is clearly marked by the change from medium-dark-gray siltstone of the Ithaca into the dark-brownish-gray shale of the Middlesex.

Depositional Environment

The sediments that make up the Genesee Formation in western New York accumulated in the eastern part of a wide and relatively shallow epicontinental sea covering much of the central United States during the Late Devonian. The sea lay west of a rising landmass that was the source for most of the detrital Genesee sediments. The large and growing Catskill Delta appended the source area and extended west into the sea. The rocks of the Genesee Formation in western New York are the distal facies of the deltaic wedge.

The abundant black, brownish-black, and very dark gray shale and mudrock in the lower part of the Genesee Formation, are believed to have accumulated in a euxinic environment, in a sea so deep and quiet that the basal waters were deficient in oxygen much of the time. Presumably the black and brownish-black shales represent deposition in relatively tranquil water with a depositional energy level just sufficient to move fine-grained quartz, silt, clay, and macerated plant detritus but insufficient to sort and winnow them into discrete beds and layers of individual components.

The sea in which the lighter gray mud and silt of the Penn Yan and the West River accumulated appears to have been slightly better oxygenated. The numerous beds of siltstone in this part of the succession are largely free from carbonaceous detritus and indicate a level of depositional energy high enough to sort and winnow the fine-grained sediments. The overall vertical decrease in darker colors from the Geneseo through the Penn Yan and West River Shale Members indicates a general clearing of the Genesee sea, although brief recurrences of euxinic conditions are shown by the many stringers and beds of brownish-black or grayish-black mudrock or shale intercalated in the Penn Yan and West River.

The Renwick Shale Member marks the presence of a widespread reducing environment in the vicinity of Seneca and Cayuga Lakes. Possibly some of the thinner and less extensive beds of black shale may indicate short episodes of extensive and prolific algal blooms and the resultant brief fouling of the sea and accumulation of abundant unoxidized plant remains on the sea floor. The relatively meager fauna in the dark mudrock and shale of the Genesee suggests an environment hostile to most benthonic forms. However, the presence of abundant *Styliolina* and other pelagic forms in the Genundewa Limestone Member and in other calcareous beds and associated shales suggests that at times the upper layers of the sea were well supplied with oxygen and nutrients.

The North Evans Limestone is a lag deposit concentrated by relatively mild scour of the uppermost beds of the Hamilton Group in western New York. During this brief episode of erosion, most of the soft recently deposited muds were swept away, and conodonts, heavy minerals, and other relatively insoluble components from the eroded strata were concentrated on a broad shoal in extreme western New York. Possibly much of the erosion took place just before and during deposition of the lower part of the Tully Limestone in central New York and before the general quiescence of early Genesee time.

The Genundewa Limestone Member suggests depositional energy sufficient to winnow the accumulating detritus and to sweep away the fine-grained muds that would otherwise dilute the accumulating mass of *Styliolina* shells that were raining down from the upper layers of the Genesee sea. The apparent lack of preferred orientation of the shells suggests that the currents were not strong enough to reorient the shells when they came to rest on the sea floor.

The turbidite beds of the Sherburne Flagstone Member and the Ithaca Member indicate incursions of high-energy turbidity currents into the relatively tranquil deeper water of the Genesee sea. Distal turbidites, which make up much of the Sherburne at Cayuga Lake and much of the Ithaca west of Seneca Lake, indicate lower level of depositional energy than do the thicker and more proximal turbidites in the area east and southeast of Cayuga Lake. The distal turbidite beds are more sheet like and lack basal channels incised in the subjacent strata. The presence of benthonic forms in local abundance and the evidence of many flow structures, including zones of flow rolls, in the upper part of the Ithaca Member in the vicinity of the Tioughnioga valley indicate the transition to proximal turbidites and a shift to a nearer shore high-energy environment.

Thus, the Genesee Formation shows an increase in depositional energy both vertically through the formation and laterally from west to east. The lateral change is considerably more conspicuous than the vertical change. Probably the most marked incursion of a turbidity flow is shown by the extent of the Bluff Point Siltstone Bed of the West River Shale Member far west of the general limit of Ithaca turbidites at Keuka Lake. The absence of other turbidites in the Penn Yan and West River Shale Members west of Keuka Lake suggests that the energy of most Ithaca turbidites was spent by the time the flows reached the vicinity of Seneca Lake and that turbidity flow ceased in the area between Seneca Lake and Canandaigua Lake. The presence of distal turbidites as much as a foot thick at Keuka Lake that are missing in the equivalent mudrock and shale at Canandaigua Lake strongly supports this conclusion.

Rocks Above the Genesee Formation Sonyea Formation

The Middlesex Shale Member of the Sonyea Formation overlies the Genesee Formation throughout the study area. The Middlesex is composed of grayish-black, brownish-black, and olive-black shale in the area from Lake Erie to Seneca Lake. Locally, discoidal to enlongate ellipsoidal silty grayish-black limestone nodules 1 foot thick and 2-5 feet long are intercalated in the basal part of the Middlesex Shale Member. Dark- to medium gray argillaceous to quartzitic siltstone is present in the member east of Seneca Lake. The beds of siltstone increase in number and thickness, particularly in the Dryden and Harford quadrangles. The Middlesex becomes silty and ranges in color from very dark gray to dark olive gray in the Tioughnioga valley area. Fragments of plant fossils are locally abundant in the member in the vicinity of Marathon and Glen Aubrey. Flow rolls, small scour channels, and complex cross laminations are common in the siltstone intercalated in the Middlesex east of Ithaca, and the number and size of the features increase to the east and are abundant near Glen Aubrey.

Chapter 4. Watkins Glen State Park Geology

Woodrow (no published date) provided a description of the Watkins Glen State Park geology as part of the geological field trip into the region.

Park Geology

Vertical cliffs about 50 m high bound the Schuyler Creek draining from the narrow mouth of Watkins Glen Falls. In the cliffs exposed medium gray, fine-grained sandstones and dark gray, silty shales of the Penn Yan (and West River?) Shales of the Genesee Group prevail. (**Figure 14**).

Figure 14. Watkins Glen Falls exposes the Genesee Formation Penn Yan and West River shales in the cliff walls. Source: Loaded Landscapes, posted on the internet.

With the Geneseo below, these rocks and the younger Devonian rocks above form a basin-fill about 2000 m thick in this region, of which all but a tiny percentage is siliciclastic. Exposures of the Group are available at many other locations around the south end of Seneca Lake. These rocks extend, at approximately the same elevation, from north of Watkins Glen south to Montour Falls village where they are exposed in a falls. The Fir Tree Point anticline crests about 4 km north of Watkins Glen and this fold disrupts the regional southerly dip of a degree or less and renders the sequence nearly horizontal for a north/south distance of at least 10 km.

Watkins Glen village has long been the site of salt production and tourism and, occasionally, disastrous flooding. Two companies are presently extracting salt by solution from the Salina Group about 700 m below ground. Tourists come to WGSP and its campground, the Glen auto racing track on the hill above, the nearby wineries, and Seneca Lake. Few people, however, are aware of the flooding potential posed by the creek which flows through WGSP, Watkins Glen village and into the adjoining marsh. The stream flows out of the Glen on to the alluvial fan on which the village is built. This means that much of the village is subject to the sedimentary processes associated with fan development.

An example: on a summer evening in July, 1938, 7-8" of rain in a few hours time resulted in a mass of water, mud, boulders, tree trunks, pieces of railroad bridge and other flotsam plugging the Glen mouth and then bursting out on to the town. In the resulting flood, one person died, several houses were destroyed, and much of the village suffered damage. Major flooding events look to have about a high recurrence at intervals of about 30-50 yrs and are little thought about in the interim.

The sandstones and shales in this exposure are arrayed in cm-scale fining-upward sequences. Body fossils are extremely rare but there is scattered plant debris and some bioturbation. The sandstone beds are sharp-based with their basal surfaces marked by groove casts and rare, flutes. The sandstones are fine- to medium-grained at the base, some contain shale chips, and they fine upwards to siltstone and shale. Well defined small-scale and steep cross-strata, some arrayed as climbing ripples, are found at the base of individual beds with larger-scale, low angled cross strata at the top. Sandstones grade into shales at the top of each couplet. The sequence looks to be a stack of turbidites made by dilute, relatively slow-moving and lightly erosive density currents. No complete turbidite sequences are found in these exposures (**Figure 15**). Most are based on Bouma subdivision "c" and a few are based on subdivision "b". The Bouma sequences are defined in the following graphic (**Figure 16**).

Bouma Sequence Discussion

Blatt and others (1980) provided an introduction to the Bouma sequences described in this section's introduction.

Introduction

In some sequences such as the ideal sequence of sedimentary structures observed in turbidites, certain structures are missing or replaced by others. In a Bouma sequence, there is a massive division below the plane of lamination that is, in turn, followed by a small scale cross bedding. The large scale cross bedded unit is absent. The Bouma sequence fits an interpretation in terms of gradually reducing flow regime.

Figure 15. Gravelly beds underlie thicker sequences of sandstone which grade upward into thinly bedded shale sequences. Although Woodrow claims there is an absence of turbidites deposits, this exposure appears to disagree with the assessment.

The exceptions stimulated additional research and sequences and are now described where large scale cross bedding is present at the expected position in the sequence. Convincing examples of antidune structures were also described from the base of some turbidites. Problems continue to remain, but the ability of the concept is present.

There are some advantages to use to concept of flow regime rather than simple hydraulic variables such as velocity, shear stress, or stream power. The flow regime concept emphasizes the fact that the assemblages of characteristic bed forms are controlled not by a single hydraulic variable but by a complex of variables. In most cases, the geologist does not know which one of the many possible hydraulic variables is responsible for a change in sedimentary structures observed in a stratigraphic section so it is not possible to identify the hydraulic variables except to say that they combined to give the conditions necessary for the development of a certain flow regime.

Flow Regime Indicators in Watkins Glen State Park

It should be noted that the sequences in and near Watkins Glen State Park are typical of those seen in the facies of Late Devonian of New York and Pennsylvania which strongly suggests that the Late Devonian sea floor in this region was of low, steady gradient. Abrupt changes in gradient were the exception, not the rule. This interpretation is based on the absence of channeling, slump scars, slumped masses and thick "slope muds," all typical of modem clay continental slopes. Channels are rare in this sequence demonstrating there was only a single channel-fill in rocks of this facies known and that are exposed along 1390 south of Danville, NY. Further, sediment transport and deposition on a low gradient ramp would explain the weakly erosive turbidity currents responsible for the turbidites seen here (see Figure 15 for turbidity deposits). As differentiated by color in the cliff face, it is apparent that there are 3-5 m thick repetitive sequences, probably fifth-order sequences or cyclothems (**Figure 17 & 18**).

The following photographs demonstrate the type of sediments that are associated with the graphic provided in Figure 16. Starting with the conglomerate evidence produced in Figure 15, this would correspond to Bouma sequence "A". It is thought that this sequence represents the turbidite layer typically found at the base of the cyclothemic sequence proposed for the Penn Yan-White River members of the Genesee Formation in the park, known as the Sherburne Flagstone Member. This sequence designation represents an incursion of high energy turbidity current into a tranquil deep water of the Genesee Sea. The Sherburne Flagstone Member is a distal turbidite that is it formed closer to the leading edge of the Catskill Delta which exhibits a lower level of depositional energy and is more sheet like, lacking basal channels. Most of the energy depositing this unit was spent by the time the flow reached Seneca Lake, north of Watkins Glen.

Sandstone deposited on top of the turbidity flows (Figure 15) and in **Figure 17** represents Bouma sequence "B", a plane parallel depositional sequence subjected to continued upper flow regime where the flow was beginning to become reduced in energy levels. Flow regime began to change into quiet water deposition, the lower part of a low flow regime where small scale ripple marks are evident in the thick Penn Yan-White River member shales. Ripple marks are visible and may be observed where the shale has been eroded into flat sheets, usually in the stream channel when water flow is low during the summer months.

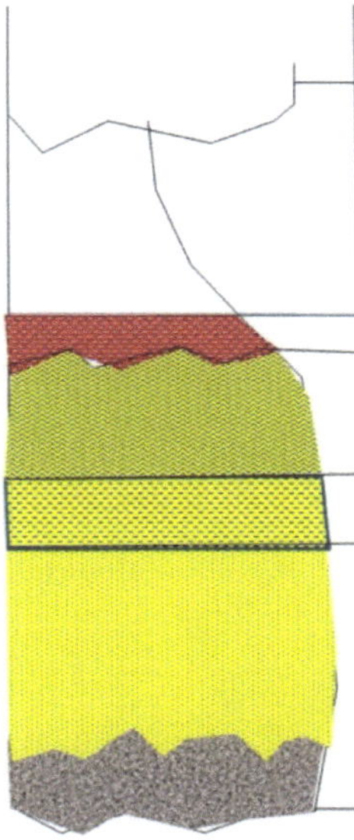

Grain size		Bouma Divisions	Interpretation
medium	E	Pelite	Pelagic sedimentation or fine grained, low density turbidity current deposition
silt	D	Upper parallel laminae	?????
sand	C	Ripples, wavy or convoluted laminae	Lower part of low flow regime
sand	B	Plane parallel laminae	Upper flow regime plane bed
Sand to granule at base	A	Massive, graded	Upper flow regime rapid deposition and quick bed (?)

Figure 16. Bouma sequences for sedimentary rocks.

Bouma "C" sequences are present in **Figure 18** where massive shale makes up the canyon walls. Deposition was in deeper quiet waters where the flow regime was nearly absent and winnowing of particles was lacking. Waters of this depth were typically euxinic, lacking oxygen which prevented benthic organisms from thriving though plants and other flora may have been flushed into the fine sediment.

Silt appeared above the shale which places the Bouma sequence at "D" which has not been mentioned by Woodrow (___) in his review of Watkins Glen State Park. Silt exposures are thicker than the shale beds but thinner than the sandstone beds in the park. Many of the bedding planes exhibit ripple marks indicating that there was enough flow regime energy to generate currents which influenced deposition (**Figure 19**).

In addition, some parts of the shale unit exhibit ripple marks which also present some interesting interpretations on the flow regime that was present at the time of deposition. Small ripples suggest there was a transition occurring between the upper flow and lower flow regimes during shale accumulation (**Figure 20**). Furthermore, the photo presents beds that are dipping more steeply to the left than those below it. This may suggest that deposition of the upper layer compacted and differentially displace the lower bed into a lesser tilted configuration; the area of deposition may have been such that the delta fore sets were deposited on top of flat lying beds which were later tilted due to compactions of the upper sediments.

Figure 21 demonstrates a view of the thick pile of thinly bedded shale that accumulated in the Catskill Delta during the Genesee Sea.

Figure 17. Sandstone beds underlie thinner beds of shale in the right wall of the canyon. Disintegrated shale is in the left canyon wall in this photo. Sandstone represents a diminishing but high energy flow regime which was reduced to quiet water when the overlying shales were deposited.

Figure 18. Repetitious deposition of shale beds line the Watkins Glen canyon supporting the observation of cyclothemic deposition during the Devonian.

Figure 19. In this photo, there is evidence of interrupted deposition in the right and left walls about half way on the cliff face. A period of non-deposition occurred which separates the lower half from the upper half of the exposure. Ripple marks are present between bedding planes and at the bottom of the creek channel. Source: Trip Advisor posted on the internet.

Figure 20. In the lower left wall above the lowermost waterfall (bracketed zone), beds appear to dip into the ground at a steeper angle than those occurring below it.

Figure 21. Cyclothemic deposition of the shale sequence is evident in this photo Source: Fifi & Hop posted on the internet.

References

Blatt, H., Middleton, G., and Murray, R. 1980. Origin of Sedimentary Rocks, 2nd Ed. Prentice Hall, Englewood Cliffs, NJ.

Bloom, A.L., 1965. Geomorphology of the Cayuga Lake Basin. Department of Geological Sciences and Institute for the Study of the Continents, Cornell University, Ithaca NY 14853 and references contained within.

DeWitt, Jr., W., and Colton, C.W., 1978. Physical Stratigraphy of the Genesee Formation (Devonian) in Western and Central New York: Stratigraphy and Conodonts of the Genesee Formation (Devonian) in Western and Central New York. US Geological Survey Professional Paper 1032-A, US Department of Interior, Washington DC.

Guild, G. 2011. The Grand Story of Time told in the Beauty of Today: The Geology of Western and Central New York. Published on the internet as a blog.

Woodrow, D.L.,___. The Late Devonian Clastic Wedge in Central New York and Northern Pennsylvania. Hobart and William Smith Colleges, Geneva, NY. Published on the internet.

www.ingramcontent.com/pod-product-compliance
Lightning Source LLC
Chambersburg PA
CBHW040246220526
45473CB00001B/391